The Art of 3D Printing

Unleashing Your Creativity in the Digital Age

Table of Contents

Chapter 1. Introduction

Immerse yourself in the transformative intersection of technology and creativity with the Special Report, "The Art of 3D Printing: Unleashing Your Creativity in the Digital Age." Within its pages, you'll journey through the groundbreaking world of 3D printing, a revolution that's reshaping everything from artistry to industry. While the topic could seem intimidating with its blend of digital and mechanical know-how, our report distills the complexity in an easy-to-understand and, crucially, a practical way. We aim to inspire not only by covering the technical nuts and bolts but also by showcasing the innovative, jaw-dropping masterpieces crafted through 3D printers worldwide. Imagine – after reading this report, you too could be taking the first steps towards expressing your unique creative vision in three dimensions. Intrigued much? Let's dive into the fascinating realm of 3D printing and unleash the budding creator in you!

Chapter 2. A Brave New World: An Introduction to 3D Printing

Let's imagine a world – not too different from our own – where each household has the ability to create objects at will. Lost a button on your favorite coat? Print a new one. Is the handle broken off your mug? Design and print a better, more ergonomic one. Or perhaps you've always desired a sculpture, one that no store carries because it originates from your imagination. With 3D printing, these scenarios are no longer the stuff of dreams.

2.1. Redefining 'Impossibility'

3D Printing, also known as additive manufacturing, is a technology that has been seeping into the mainstream, crumbling the barriers of impossibility with each layer of material it adds. At its core, 3D printing involves the creation of objects through the accumulation of thin layers of material, typically plastic, metal, or ceramic. Using a digitally created model as a blueprint, 3D printers extrudes layer upon layer of material until the desired object takes shape.

Do not, however, look at 3D printing as a mere tool. It is an invention akin to the hammer that drove the first nail of the house, the pen that signed the peace treaty, and the camera that captured our triumphs and adversities. Like these pioneering elements of human progress, 3D printing isn't just changing something; it's changing everything.

2.2. The Dawn of 3D Printing

Unbeknownst to many, the concept of 3D printing has actually been around for a few decades. It was the brainchild of an engineer named Chuck Hull who, in the 1980s, invented a process he called 'stereolithography'. He aimed to create physical objects from digital data, a concept that was revolutionary at the time.

The technology, however, remained confined to the realm of engineering and large-scale manufacturing due to high costs and a commanding level of technical sophistication. Yet, as the digital age gained momentum and technology became more accessible, so too did the concept of turning digital 3D models into tangible reality.

Now, as breakthroughs occur in fields from biotechnology to architecture, 3D printing increasingly encourages us to question traditional limitations. The promise of personal manufacturing, customizable products, and even life-saving applications in medicine has broadened our conception of what's possible.

2.3. The Mechanics of 3D Printing

What grants 3D printing its transformative potential is its core process: additive manufacturing. As the name implies, it's all about 'adding'. Traditional methods for creating objects often involve subtracting or carving away material, often resulting in wasted resources.

3D printing, however, carefully adds material layer by meticulous layer, only utilizing what's necessary for the object at hand. Each printed layer adheres to the preceding one, building your dream object slice by thin slice, much like a loaf of bread.

2.4. Decoding the 'Digital' in Digital Fabrication

3D printing is a type of digital fabrication – and understanding the 'digital' in this equation matters. Digital fabrication reverses the traditional process of manufacturing. Instead of going from tangible materials to a final product, it works from a digital interface to manifest the physical item.

Any 3D printing journey begins with a 3D model. These models are created on a digital platform employing Computer Aided Design (CAD) software or scanned from an existing object. The resulting digital file is then sliced into thin cross-sections by the software — a method any printer would recognize as 'preparation for printing'.

The file's translation into a language the printer understands, typically G-code, informs the 3D printer about the exact specifications, layer thickness, patterns, and the sequence of printing to craft the object accurately.

2.5. Applications of 3D Printing: Stretching the Boundaries

3D printing has a tremendous potential to usher in positive changes in a plethora of fields. Some are familiar – like using 3D printers to manufacture consumer goods, car parts, or architectural models. Others are quite extraordinary, pushing the boundaries of what we can conceptualize.

A promising area that is steadily gaining traction is bio-printing. By utilizing materials mimicking human tissue, scientists aspire to 3D print organs to address the shortage encountered in transplants. Imagine the life-changing (or rather, life-saving) potential that this branch of 3D printing nurtures.

Equally remarkable is constructing buildings through 3D printing. The world's first 3D printed office in Dubai, for instance, heralds an era of efficient, environmentally friendly and potentially more affordable construction.

From art to fashion, education to space exploration, a diverse tapestry of fields can utilize 3D printing to reduce waste, drive efficiency, encourage creativity, and enhance the overall qualitative experience.

We are standing at the precipice of a new era, one marked by an exponential explosion in creativity and innovation — all thanks to 3D printing. This is just the introduction, the first page - the technology is still in its infancy, yet poses the promise of an exciting future where everyone can be a maker and creator. As we navigate through the pages of this new chapter, it will be fascinating to observe and participate in the evolution of 3D printing.

Chapter 3. The Mechanics of Dreams: Understanding 3D Printing Technologies

The world of three-dimensional digital fabrication, more commonly known as 3D printing, is a vast landscape of technological innovation packed with the potential to redefine both art and industry. The transformative power of this technology lies in its ability to construct complex and precise physical objects from digital blueprints - turning dreams into reality.

To comprehend the awe-inspiring world of 3D printing, we first need to acquaint ourselves with the foundational technologies underpinning this revolution. Don't let the jargon intimidate you. Our mission is to make you thoroughly understand the process as effortlessly as sipping your morning coffee.

3.1. Embracing the Three Dimensions

In elementary terms, a 3D printer constructs objects layer by layer, ranging from simple geometric forms to complex scaled models. This is a stark contrast to other manufacturing processes; often involving the removal or alteration of existing material, like carving a sculpture from a piece of marble.

The journey from a digital design to a tangible object starts with a 3D model created in a Computer-Aided Design (CAD) program, or obtained from a 3D scanner or modeling software. The digital model is then "sliced" into very thin layers, each equivalent to one layer of the printed object. These slices guide the printer's movements, allowing it to lay down precise amounts of material layer by layer.

This process abides by the Cartesian coordinate system, a mathematical concept we've all grappled with in school. The 'X', 'Y', and 'Z' axes denote the three dimensions within which the printer operates.

3.2. The Materials: More Than Just Plastic

While plastic, in various forms, is indeed the most common material used in 3D printing, the technology's ever-growing ambit includes ceramics, metals, sand, and even biological substances.

Plastic often comes in the form of Filament, Resin, or Powder. Thermoplastics like ABS (Acrylonitrile Butadiene Styrene) and PLA (Polylactic Acid) are the most commonly used filaments. On the other hand, Powder & Resin printers use a laser or ultraviolet light, respectively, to solidify material.

Metals, used in Direct Metal Laser Sintering (DMLS) and Selective Laser Melting (SLM) technologies, open up a broad avenue to industries like aerospace, automotive, and medical device manufacturing.

In Bioprinting, a mixture of cells and biocompatible materials printed layer by layer can create tissue-like structures. It's a promising avenue for regenerative medicine.

3.3. Unraveling the Printer Types

Let's explore four fundamental types of 3D printing technology, each with its unique approach to creating 3D prints.

1. Fused Deposition Modeling (FDM): The most popular 3D printing technology, FDM melts thermoplastic filament and extrudes it through a nozzle layer by layer. This process is akin to using a hot

glue gun moved along predetermined paths.

2. Stereolithography (SLA): This technology uses a vat of curable photopolymer resin. An ultraviolet laser is strategically moved across the resin's surface, hardening it wherever it touches, forming thin layers that constitute the final product.

3. Selective Laser Sintering (SLS): Instead of liquid resin, SLS uses powdered materials, such as nylon or metal. A laser sinters the powder together to form a solid structure.

4. Direct Metal Laser Sintering (DMLS): A type of SLS, DMLS is specifically tailored to work with metal powders.

3.4. All About the Resolution

3D printer quality, or resolution, plays a significant role in determining the fidelity of the final output in terms of intricate detailing and surface finish. It is generally measured in microns, with FDM models having a resolution range from 100 to 200 microns. SLA and DMLS, thanks to their precise lasers, can achieve resolutions up to 25 microns, making them suitable for parts demanding high detail.

3.5. Software Matters

The 3D printing process isn't merely a hardware play. Software applications, like CAD tools for designing, slicer software for defining the printing strategy, and firmware to operate the printer, work in unison to create the final output. These software tools allow for pre-print customization, ensuring an accurate translation of your creative vision into a 3D object.

Embracing the diverse world of 3D printing demands an understanding of its multifaceted aspects. The lifecycle of a 3D printed object—from digital model to tangible output—presents us with a blend of art and engineering, a testament of human ingenuity and creative prowess. With our insights on its foundational technologies, we hope this chapter helps to demystify 3D printing and encourages you to step into this remarkable realm.

Chapter 4. From Design to Reality: 3D Modelling & Software Tools

3D printing leaps from concept to reality through the use of sophisticated design tools and software. This chapter will unveil the mystery behind this intricate process, making it accessible and simpler for your budding creativity.

To better understand how a design goes from a mind's idea to a physical creation, let's demystify the process it undergoes. Our journey starts with 3D modeling, where you render your ideas into tangible models.

4.1. The Art of 3D Modeling

3D modeling breathes life into concepts. Using computer graphics, 3D modeling enables you to construct your design digitally, creating a representation of your idea's physical characteristics. The techniques used are broadly categorized into two main modeling types: Parametric and Freeform.

Parametric modeling uses parameters, such as dimensions, to define a model accurately. It provides high precision and comes in handy when you're working on complex, precise models. Freeform modeling, on the other hand, is sculpting your idea in a freeform way; it's perfect for creating organic models with more flexibility in shape and design.

For beginners looking to dip their toes in the 3D printing world, tools such as TinkerCad offer an easy entry point into 3D modeling. For those with more experience or ambition to create complex models, software such as AutoCAD, SolidWorks, and Blender can deliver high-detail output.

4.2. Choosing the Right 3D Modeling Software

The evolution of 3D printing technology has led to an array of 3D modeling software catering to different needs, allowing you to create everything from simplistic shapes to intricate designs. But which software should you choose?

Consider the following factors while choosing your 3D modeling software:

- Proficiency level: Beginner, Intermediate, or Expert.

- Purpose: Whether you're designing for fun, creating models for commercial use, or engineering related designs.

- Budget: Some powerful tools come at a premium, but there are also free and open-source options available.

- Platform: Does your computer system support the software?

To lighten the decision-making burden, here's a list of popular software used for 3D printing along with their selling points.

1. Tinkercad: Perfect for newbies, this software is free and offers an intuitive user interface.

2. Blender: Known for its versatility, Blender is a free and open-source software ideal for intermediate to advanced users.

3. SolidWorks: This premium software caters well to engineers and professionals, known for its robust capabilities in parametric modeling.

4. AutoCAD: An industry standard for precision and professionalism, AutoCAD is widely used by architects and engineers.

5. Fusion 360: Offering a blend of parametric and direct modeling, this software is great for hobbyists and professionals alike.

4.3. From 3D Modeling to 3D Printing: Transitions and Translations

Once your 3D model is ready, the next step is to translate it into a language that your 3D printer understands: G-code. This process of conversion is achieved through software programs known as 'slicers.'

Slicing software takes your 3D model and chops it into layers that your 3D printer will sequentially print to build your finished object. Each layer corresponds to a precise movement path your printer's nozzle will follow as it lays down material.

Most 3D modeling software offers integrated slicing capabilities, but standalone slicers like Cura, Simplify3D, and Slic3r can also be a powerful option. These applications not only generate the G-code but also allow you to modify the settings of your 3D print – for instance, print speed, layer height, and fill density.

Starting from a simple idea, we have journeyed through 3D modeling software and slicers until finally, we are ready to bring your creation to life. With one more click, your 3D printer will start moving and laying down material layer after layer – bring your idea from the digital realm to a physical reality.

As you explore these tools, remember: it's not just about the technology, but about how you interpret it, master it, and apply it to unleash your creative potential in 3D printing. Now venture forth, empowered, to craft your tangible dreams!

Chapter 5. Raw Creation: An Overview of 3D Printing Materials

One of the most transformative aspects underpinning the world of 3D printing is the vast array of materials available for creation. The choice of material not only influences the quality of the print but also its durability, flexibility, detailing, and even its color. This chapter would dive into the different types of materials available for 3D printing, their properties, and their potential uses in various applications, from everyday items to intricate artistic pieces.

5.1. Plastics – The Basic Building Blocks

Plastics are the most commonly used materials in 3D printing due to their versatility and accessible price point. The two primary types of plastics used in 3D printing are ABS (Acrylonitrile Butadiene Styrene) and PLA (Polylactic Acid).

ABS is robust, flexible, and heat-resistant, making it ideal for creating functional parts such as gears, connectors, and all sorts of durable components. However, ABS does produce fumes during printing, so proper ventilation is needed when working with it. It also requires a heated print bed to prevent warping, adding a level of complexity to the 3D printing process.

PLA, on the other hand, is a biodegradable material derived from renewable resources like cornstarch or sugarcane. PLA doesn't produce harmful fumes, so it is safer to use in a home or school setting. It offers a lower melting point compared to ABS, meaning it doesn't require a heated print bed, which again simplifies the printing process. However, PLA can be more brittle than ABS, which is something to keep in mind depending on the application.

5.2. Advanced Plastics – For the Sophisticated Artisan

Beyond the basics, there's a slew of advanced plastic materials, including PETG, Nylon, and TPU, each with unique properties that open up new possibilities for 3D printing.

PETG (Polyethylene Terephthalate Glycol) is known for its clarity, flexibility, and impact resistance, and it's FDA-approved for food contact. It can be a bit tricky to print with, however, it's excellent for applications that need both transparency and flexibility.

Nylon is a highly durable and flexible material with high heat resistance. It's commonly used in mechanics and engineering for creating resistant mechanical parts.

TPU (Thermoplastic Polyurethane) is a flexible, rubber-like material. It's highly flexible and resilient, making it perfect for creating products that can withhold stretches, twists, and bends.

5.3. Metals – The Frontier of Durability and Precision

The use of metals in 3D printing, although not as widespread as plastics due to its cost, offers immense possibilities in terms of strength, durability, and detail. The most commonly used metals include stainless steel, titanium, and even precious metals such as gold and silver.

Stainless steel is used for parts that require both strength and corrosion resistance, while titanium, with its extraordinary strength and light weight, is widely used in aerospace and medical applications.

Precious metals, such as gold and silver, have found a place in high-end jewelry design. They give jewelers the ability to create intricate designs that would be otherwise impossible or prohibitively expensive to produce.

5.4. Ceramics – The Artistic Touch

Ceramics in 3D printing have increasingly gained popularity, especially in artistic fields and home decor for their unique aesthetic qualities. They produce stunning pieces that are heat resistant, food safe, and can be glazed and fired just like traditional ceramics.

5.5. Exotic and Experimental Materials – Daring to Dream

The range of materials doesn't stop there - from wood to conductive filaments, glow-in-the-dark plastics, and even materials infused with metals or carbon fiber, the possibilities are limited only by your imagination.

Wood or Wood-fill filaments are a mixture of PLA and finely ground wood, giving the printed parts a real wooden look and feel. These are excellent for decorative items and toys.

Conductive filaments are plastics infused with conductive materials, opening the doors for printing low-voltage electronic circuits and touch-sensitive devices.

Glow-in-the-dark plastics are PLA or ABS filaments with phosphorescent material, mainly used for novelty or decorative prints, adding a fun element to your creations.

Finally, on the experimental front, we have exciting developments like carbon fiber-infused or metal-infused filaments, which are pushing the boundaries of what is possible with 3D printing.

In conclusion, the world of materials available for 3D printing continues to expand and evolve, opening new doors for innovation and creativity in various fields. The potential is as boundless as the imagination of the designers and inventors who use these materials to bring their visions to life. With a good understanding of the types, properties, and potential applications of 3D printing materials, you are well on your way to making informed decisions about your projects, and maybe even contributing to the ever-growing possibilities in the world of 3D printing.

Chapter 6. Onwards Towards the Future: Advanced 3D Printing Techniques

3D printing, albeit a modern miracle in many respects, remains firmly grounded within its foundational techniques - fused deposition modeling (FDM), stereolithography (SLA), and selective laser sintering (SLS), among others. Yet, like all technology, it persists in an evolutionary narrative, constantly seeking and adapting to new advancements. As we look towards the future, we encounter sophisticated techniques challenging the status quo of what's possible within the sphere of 3D printing.

6.1. Introducing Multi-Material 3D Printing

With the advent of multi-material 3D printing, the one-size-fits-all notion of printing material is swiftly becoming a relic of the past. Single material prints, although immensely transformative in their right, lack a level of complexity that multi-material printing readily offers.

Multi-material 3D printing allows the simultaneous use of different materials in a single printed model. By incorporating a selection of materials into the printing process, we can realize objects that not only mimic the visual characteristics of the intended design but echo its varied tactile and functional properties as well. With multiple materials at its disposal, the 3D printer can create outputs that are multi-colored, multi-textured, and even multi-functional, engaging numerous mechanical properties like elasticity and rigidity in a single model.

Imagine a 3D-printed smartphone casing that's resilient on the outside, adorned with soft, cushion-like buttons, fashioned in multiple colors - all produced in one single run. This is the promising future that multi-material 3D printing ushers in.

6.2. Expanding Horizons with 4D Printing

If 3D printing revolutionized manufacturing, then 4D printing promises to alter that revolution altogether. It is here where the printed object takes on a life of its own, dynamically responding to specific stimuli such as temperature, light, or even magnetic fields.

4D printed objects are first printed in a conventional 3D manner, albeit with specially programmed "smart" materials. These materials possess latent qualities that are activated under the intended stimuli, compelling the object to evolve and reshape.

It's not all science fiction. Imagine pipes that expand or contract based on water pressure changes, bridges that can self-assemble in hostile environments or medical implants designed to change shape once inside the body. The implications are countless and impact almost every field from construction to medical science to fashion, rendering 4D printing one of the most exciting advancements in the world of technology.

6.3. Nanoscale 3D Printing - The Micron Revolution

The journey into the future of 3D printing takes us not only across dimensions but also into minute realms, where creations measure mere microns. The emergence of nanoscale 3D printing – or 2-photon polymerization (2PP) – is dramatically pushing the boundaries of this technology.

2PP technique employs the use of ultra-short laser pulses to catalyze a photosensitive material's hardening. The precision is so fine-tuned that we've moved far beyond millimeters and now have control at the level of nanometers. This groundbreaking method allows us to create micro-structures with intricacies and details unprecedented in standard 3D printing.

With 2PP, we're opening up new micro-worlds in fields such as medicine, where bespoke cellular structures can be printed. It also finds purpose in the field of microfluidics, contributing to the development of labs-on-chips - a promising instrument in diagnosing diseases at an early stage.

6.4. Generative Design and 3D Printing - An Evolutionary Leap

Generative design is a technology where computer algorithms are used to synthetically evolve the most optimized design solution by running multiple iterations. When employed alongside 3D printing, it results in components and designs that were once inconceivable.

In the case of generative design, the user simply specifies system goals and constraints - such as weight reduction, load-bearing capabilities, or cost restrictions. The computer, acting like a digital 'Mother Nature,' iterates a multitude of potential solutions, selecting and evolving the most optimized designs. When finalized, these intricate and efficient designs, often impossible to manufacture using traditional methods, can be printed using a 3D printer.

From creating sleek, optimized components for aerospace applications to devising organically-inspired furniture pieces, the combination of generative design and 3D printing is crafting the future right now.

As we travel forward, the advancements only seem to multiply, opening doors to concepts like bioprinting and artificial intelligence-infused processes. What is evident is that 3D printing is far beyond a mere tool - it's an evolving art form, giving us deeper insights into the boundless intersections of technology, art, and craft. Let's embrace the wave and, as creators, ride it towards the horizon of possibility.

Chapter 7. Aesthetic Innovation: Inspirational Examples of 3D Printed Art

The advent of 3D printing technology has opened up a new vista for artists across all mediums to explore, expand, and reinvent their creative vision. This remarkable process, also known as additive manufacturing, molds materials into unique and elaborate forms, enabling artists to construct complex creations that were previously confined to the boundaries of imagination. Here we delve into the world of 3D printed art and showcase some inspirational examples that illuminate the aesthetic and innovative potential of this futuristic tool.

7.1. The Phenomenon of 3D Printing in Art

The use of 3D printing in art is not merely an adoption of a new tool, but a paradigm shift in the process of creation itself. It often involves a digital design phase, which expands the creative possibilities to incorporate designs from 3D scans, digital sculpting tools, and mathematical algorithms. The virtual designs are then actualized utilizing the layer-by-layer material addition that defines 3D printing. This novel method of fabrication gives artists an unprecedented degree of control and precision, allowing them to realize forms with intricate geometries and hollow structures that would not be possible using traditional art-making practices.

One of the exceptional attractions of 3D printing for artists is the versatility in available materials. While plastic polymers like ABS and PLA are the most common, artists are also exploring printing in a wide array of materials, including metals, ceramics, glass, and even biodegradable substances like algae and fungi. This extensive material scope gives artists the flexibility to produce works with a vast range of properties — from the strictly functional to the extravagantly aesthetic.

7.2. Pioneers Pushing the Boundaries: Key Artists in 3D Printing

There are numerous artists globally who are pushing the boundaries of 3D printing technology and redefining what is possible, creating extraordinary works that inspire, stir curiosity, and provoke thought.

One such pioneering artist is Neri Oxman, an architect, designer, and professor at the MIT Media Lab. She develops the concept of 'material ecology,' where 3D printing is used to create objects that interact organically with their environment. For example, her "Silk Pavilion" installation involved 3D printing a scaffold and then setting loose 6,500 silkworms to weave a natural coverage over the structure. The result was a stunning, semi-transparent dome that demonstrated a beautiful melding of high-tech and biological creativity.

Another artist worth mention is Nick Ervinck, a Belgian visual artist known for his vibrant and surreal structures. Ervinck often incorporates organic and mechanical elements in his designs, which he then materializes using 3D printing technology. His works, like the monumental yellow sculpture 'GNI-RI,' offer a vivid example of how the digital and physical realms can harmoniously coalesce.

7.3. Art Installations with a Purpose

Beyond visual intrigue, 3D printed artworks are also confronting critical global issues and instigating dialogue around them. Dutch artist Daan Roosegaarde, known for his innovative environmental installations, used 3D printing for his project 'Smog Free Tower'. The tower, designed to clean the polluted air in urban environments, incorporated 3D printed components to provide structural durability as well as aesthetic appeal.

Likewise, the Mexican artist Gilberto Esparza has produced bio-artistic pieces, such as his 'AutoPhotosynthetic Plants,' that incorporate 3D printed elements. Esparza's work explores the relationship between biological organisms and technology, presenting thoughtful analysis of ecology in the digital age.

7.4. Thought-Provoking Sculptures and Abstract Ideas

Abstract ideas and concepts are frequently realized through the unique possibilities of 3D printed art. Salient examples can be found in the work of Joshua Harker, one of the pioneers of 3D printed sculpture. Harker has used 3D printing to create intricate, lace-like skulls and other objects that defy the norms of geometry and design.

Keith Brown, a UK based sculptor and professor, is another artist whose work has married traditional sculpting techniques with digital realization. Brown's complex, abstract forms push the boundaries of material and conceptual innovation in 3D printing, encouraging a rethinking of how art can exist and evolve in three dimensions.

In conclusion, the realm of 3D printed art is vast and constantly morphing. As artists continue to innovate and push the boundaries of this technology, we can expect to see even more astonishing and thought-provoking creations. The artists and works featured here represent merely a fraction of the diversity and potential harbored within this dynamic field, offering a glimpse into its rich possibilities. As this narrative unfolds, one thing is abundantly clear: 3D printing's aesthetic power is not only transforming the visual arts, but it is also challenging our perceptions, inviting us to reimagine the future of creativity.

Chapter 8. Printers at Play: Exploring 3D Printing in Toy and Game Design

3D printing has truly emerged as an efficient and innovative solution for the toy and game industry, breathing life into inventive ideas that weren't feasible with traditional methods of toy manufacturing. With additive manufacturing technology becoming more prevalent, affordable, and user-friendly, it's becoming easier for hobbyists, as well as professionals, to create their toys and games.

8.1. A Spectrum of Possibilities in Toy Making

The potential applications of 3D printing in the arena of toy making are manifold. From action figures and dollhouse furniture to intricate board games, the toolset's limit lies within one's imagination. 3D printed toys are no longer limited to simple geometric shapes, thanks to the advancement in additive manufacturing technology and digital designing software, which allow for the creation of complex designs with sophisticated details.

If you're a fan of action figures, imagine designing your superhero and seeing it come to life. 3D printing enables toy enthusiasts to personalize action figures, making them more distinctive and cherished. With digital sculpting tools, artists can design their action figures in a 3D modeling software and print them layer by layer in a variety of materials such as plastic, resin, or even metal.

For dollhouse enthusiasts, the 3D printer can transform an empty dollhouse into a fully furnished home, complete with Victorian-style miniature furniture, tiny kitchenware, and petit picture frames.

There are numerous online platforms available where 3D printable files for miniature furniture are shared by the community. One can easily find a design they love, download the file, and print it.

Using 3D printing, fans of board games can not only replace lost pieces but create entirely new games. With software such as Tinkercad or Fusion 360, one can design the game pieces, while printable sand or silicone molds can be created for mass production of identical pieces.

8.2. Challenges and Limitations

Although 3D printing offers a plethora of opportunities, it is not free from challenges. The technology enables the quick transformation of digital models into physical objects, but the process requires adequate knowledge of 3D modeling or CAD software.

Material and durability are other significant considerations. While plastic is an accessible choice for many home 3D printing enthusiasts, it may not offer the desired durability for certain toys, particularly for those handled roughly or frequently by children. Other materials like metal or resin offer more durability but come at a higher cost and require advanced printers.

Another challenge particularly important in the sphere of toys and games is compliance with safety standards. Toys, especially those intended for younger children, need to meet certain safety norms related to material toxicity, size, and durability to prevent harm to end-users.

8.3. Industrial Application and Mass Production

The impact of 3D printing extends beyond hobbyists and extends to the industrial arena. Big toy companies are adopting 3D printing techniques for the prototyping process, reducing product development time and enabling rapid iterations based on feedback.

Developing prototypes through 3D printing before moving to mass production ensures that the final product resonates with the target market's needs and expectations. This approach also avoids the costly and time-consuming process of modifying molds, especially for complex designs.

In terms of mass production, 3D printing provides an efficient method to produce customized products or limited editions, a demand frequently expressed by the toy and game community.

8.4. The Future of Toy Making

As 3D printing technology continues to develop, offering increased print speed, improved print quality, and vast material ranges, we can predict that its influence on the toy and game industry will increase.

For the individual consumer, the future could see an even more direct role where they can - besides personalizing toys – design, print and modify them right at home. Companies might embrace a business model where consumers purchase digital design files and print the toy at home, leading to an entirely new commerce and digital copyright landscape.

On the commercial front, larger companies will likely further their use of 3D printing for prototyping and limited-edition releases. Though traditional mass production techniques are far from being obsolete, having 3D printers as part of their manufacturing arsenal provides companies with the flexibility to cater to varying consumer demands.

In conclusion, 3D printing is an increasingly potent tool for the toy and game industry, offering unparalleled creative freedom, reduced costs, and efficient design iterations. However, like any tool, its effectiveness depends on the user's skill level, understanding of the printing process, materials, and safety implications. The rapidly advancing technology promises an even brighter future for 3D printing in toy and game design, hurdling over its current limitations, and catering to both the avid hobbyist and the manufacturing behemoth, bringing imagination to life.

Chapter 9. Building the Unimaginable: Applications in Architecture and Construction

Before delving into the profound ways 3D printing is changing the fabric of architecture and construction, it is essential to understand its process. 3D printing, also known as additive manufacturing, is the sequential layering of materials directed by digital design data to build three-dimensional structures. This technology, although sounding futuristic, has already been adopted into numerous industries, including the medical field, automotive industry, and food industry, to name a few.

9.1. The Emergence in Architecture and Construction

Today, 3D printing technology is forging a revolutionary path in architecture and construction, reshaping the industry's way of conceptualizing, designing, and constructing buildings. It is considered a transformative tool, driving design freedom, waste reduction, cost-effectiveness, and construction speed. Such a technological advancement appears to be a remedy for the growing demand for sustainable and efficient building methods in today's fast-paced world.

The most significant benefits of utilizing 3D printing in architecture and construction are abundant, and some of them are mentioned as follows:

1. It drastically reduces waste: Traditional construction methods

involve substantial material waste. However, with 3D printing, the materials are used more efficiently because the printers only use the exact amount of material needed to construct the design. This contributes to a greener, more sustainable approach, advocating for appreciable resource savings and environmental benefits.

2. Boosts design freedom: 3D printing paves the way for architects to conceptualize designs that would have previously been considered impossible or unthinkable due to the constraints imposed by traditional construction methods. Now, intricate design elements, non-linear forms, and complex geometries can be seamlessly created, pushing the boundaries of conventional architecture.

3. Time and cost efficient: 3D printing can significantly lower expenses by shortening the construction timeline and reducing the manpower required. Printers are able to operate continuously without requiring breaks, unlike human labor, and parts can also be produced off-site and transported for assembly, further speeding up the construction process.

9.2. Case Studies of 3D Printing in Construction

To better grasp the extent of this technology's potential, let's explore some real-world applications of 3D printing in architecture and construction.

1. Project Milestone, Eindhoven, Netherlands: In 2019, a Dutch consortium began the construction of the world's first 3D-printed concrete homes that were not just prototypes but actual, habitable houses. Not only did the project deliver on-time and budget, it also showcased how 3D printing could be used for intricate detail and unconventional shapes, challenging the norms of box-like concrete construction.

2. YHNOVA "BatiPrint" Project, Nantes, France: In 2018, a consortium led by the University of Nantes built a house for a local family using a 3D printer in just 54 hours. The project encapsulated the concept of speed in construction, while also demonstrating an appreciable reduction in production costs.

3. TECLA, Italy: In 2021, the TECLA project showcased the possibilities of constructing a sustainable, low-cost house entirely from 3D-printed clay. Apart from the more common concrete, this project demonstrated the flexibility of 3D printing with different materials, expanding the horizons for eco-friendly construction.

9.3. Challenges and Opportunities

Despite its impressive applications, 3D printing in architecture and construction still faces significant challenges. To begin, many countries lack the legislation and standards to accommodate 3D-printed buildings, as this technology is still in its nascent stages. Moreover, the adoption of such tech-based solutions also requires a substantial shift in mindset among architects, builders, and clients.

However, these challenges often spur innovative solutions, and 3D printing technology offers several exciting prospects that could potentially address these problems. For instance, 3D printing can be particularly beneficial in disaster zones or under-developed areas that need quick and affordable housing. The technology can even extend to lunar or planetary construction for future space missions, as demonstrated by NASA's 3D-Printed Habitat Challenge.

As we look forward to a future characterized by environmental concerns and a constant quest for efficiency, the potential role of 3D printing in architecture and construction only becomes more significant. This new age of construction, fueled by 3D printing, is no longer a distant dream, but an emerging reality, capable of building the unimaginable.

Chapter 10. In Focus: Interviews with Groundbreaking 3D Artists and Innovators

In the colorful and endlessly evolving vanguard of 3D printing, there's an array of talented individuals whose curiosity, determination, and unbounded creativity push the boundaries of this medium. This section provides a glimpse into the minds and creative processes of five of these exceptional innovators, giving us insight into how they leverage 3D printing technology to craft their distinctive, mesmerizing works of art.

10.1. The Architectural Innovator: Taro Shinoda

Hailing from Tokyo, Taro Shinoda is renowned as a contemporary artist who has utilized 3D printing technology to reimagine modern architecture.

Q: **Could you share what triggered your interest in 3D printing?**

A: I've always had a fascination with architecture, more specifically, how it interacts with human behavior and perception. When I came across 3D printing, I saw a revolutionary opportunity to explore these aspects in completely new ways.

Q: **How do you combine traditional architecture principles with 3D printing in your work?**

A: What's exciting with 3D printing is the potential for a hands-on, creative process. It's easy to play with the scale, design minute structural details, or even build complex shapes that traditional methods can't reproduce.

10.2. The Bio-Designer: Alejandra Salinas

Alejandra Salinas, a New York-based biodesigner and artist, innovatively merges her knowledge in biology with 3D printing to realize intricate bio-artworks.

Q: **What sparked your interest in combining biology with 3D printing?**

A: As a biodesigner, I am intrigued by natural patterns and structures. 3D printing technology opened up the ability to mimic those biological forms and structures. It's an engaging way to highlight the synergy between nature and technology.

Q: **How do you think 3D printing contributes to bio-art?**

A: In bio-art, the ability to print living cells, tissues, and potentially even organs, presents almost limitless possibilities. 3D printing makes it possible to create structures that can modify their behaviors in response to environmental stimuli, just like natural organisms.

10.3. The Wearable Art Maestro: Jiachen Xu

Chinese designer Jiachen Xu has been winning acclaim for his unique approach to creating wearable art pieces through 3D printing.

Q: **How has 3D printing influenced your approach to designing wearable art?**

A: The material possibilities and precision offered by 3D printing technology has profoundly expanded the boundaries of wearable art design. It has allowed me to bring the impossible into tangible existence.

Q: **Could you elaborate on how you incorporate 3D printing in your design process?**

A: In essence, I consider 3D printing as a tool to facilitate my design. It allows me to create highly detailed, intricate pieces that capture movement and fluidity; something hard to achieve with traditional methods.

10.4. The Recycled Materials Guru: Sayaka Ganz

Japanese artist Sayaka Ganz utilizes 3D printing technology to create her unique sculptures made entirely of recycled materials.

Q: **Can you tell us about your motivation to use recycled materials in your art?**

A: My use of recycled materials stems from the belief that every object has a life and spirit. With 3D printing, I can reshape these discarded objects into something new and beautiful.

Q: **How has 3D printing helped your process of working with recycled materials?**

A: 3D printing, especially with recycled plastics, offers me the flexibility to achieve shapes and forms that are nearly impossible to realize otherwise. It's boosted the expressiveness and impact of my work.

10.5. The Mathematical Sculptor: Brent Collins

Brent Collins, based in the United States, merges figures from complex mathematics into beautiful sculptures through the use of 3D printing.

Q: **What got you interested in creating mathematical art with 3D printing?**

A: I've been enamored with geometry and patterns. The precision, possibilities, and flexibility offered by 3D printing was a creative game-changer. With it, I can bring complex mathematical concepts to life.

Q: **How has 3D printing shaped your mathematical sculptures?**

A: Absolutely magnificently. By enabling the creation of geometrical forms and patterns that would be almost impossible to create by hand, 3D printing has breathed new life into my sculptures and enriched my artistic expression.

These interviews underline just how versatile and liberating 3D printing can be as a creative tool. Every artist's approach is unique, leveraging the technology to bring their powerful visions to life. The potential is endless – who knows, maybe after exploring this array of pioneering creatives, you might find your fingers itching to create your own 3D printed masterpiece.

Chapter 11. Your First Steps: Getting Started with Your Own 3D Printing Projects

One of the most captivating things about 3D printing is your ability to bring ideas to life. Whether it's a piece of creative art, a unique piece of jewelry, or an essential tool, 3D Printing offers an exciting world of endless possibilities. With that in mind, where does one even begin? We'll cover that and more in this comprehensive guide.

11.1. Understanding 3D Printing Basics

3D printing, also known as additive manufacturing, is a process that creates a physical object from a digital design. There are different types of 3D printing technologies and materials that you can use, but all are based on the same principle: a digital model is turned into a solid three-dimensional physical object by adding material layer by layer.

A typical 3D printing process involves these main steps:

1. Creating a 3D Model: This can be done through personalized design software or by downloading a pre-made model from the internet.

2. Slicing the 3D Model: This step transforms the 3D model into cross-sections that the printer will be able to understand.

3. 3D Printing: The printer interprets the slices and physically prints the design layer by layer.

4. Post-Processing: Depending on the project materials, you may need to clean or finish your prints to give them a polished look.

Many 3D printers use common materials like ceramics, metals, plastic, synthetic resin, and even chocolate. Choosing which materials to use will depend on the purpose of your project.

11.2. Acquiring a 3D Printer

Whether you are a dedicated artist looking to expand your craft or a hobbyist eager to explore this exciting new medium, choosing your first 3D printer is a crucial decision. It's important to consider your personal needs, budget, and the features of the printer.

Here are some vital points to ponder when buying a 3D printer:

1. Purpose: Why are you buying a 3D printer? If you're an artist, perhaps you need a model with high-resolution capabilities. If you're a hobbyist or education professional, ease of use might take precedence.

2. Price: 3D printers vary significantly in cost, from a few hundred to several thousand dollars. Higher-end machines usually offer more features and higher resolution printing, but a more affordable printer may still meet your needs.

3. Materials: Ensure the machine supports the materials you intend to use. Most printers support ABS (Acrylonitrile Butadiene Styrene) or PLA (Polylactic Acid), but some high-end printers can work with a wider range of materials.

4. Size: Think about the space you have available. Desktop printers are more compact and suitable for smaller projects. Large-format printers, on the other hand, can produce larger prints, but they also take up more space.

11.3. Learning 3D Modeling

The next thing to tackle is learning how to create the 3D models that the printer will use. There are countless 3D modeling software applications available – from professional-grade software with steep learning curves, like AutoCAD, to novice-friendly options such as Tinkercad and SketchUp.

To get you started, here are a few beginner-friendly 3D modeling applications:

1. Tinkercad: This free, online 3D modeling program is perfect for beginners. It offers user-friendly tutorials and a straightforward interface perfect for novice designers.

2. SketchUp: This software is also aimed at beginners but offers depth as you progress. The free version is quite powerful, and there's a paid version available for professionals.

3. Blender: Blender is an open-source 3D creation software that's free for anyone to use. It has a steeper learning curve, but it's a great choice if you want to eventually move onto more complex projects.

Remember, mastering 3D modeling software takes time and practice. Don't worry if you don't create a masterpiece on your first try. Keep designing and revising your work.

11.4. Preparing Your 3D Model for Printing

Once your 3D model is ready, you can't simply send it directly to your printer. It must first be processed by a piece of software called a "slicer," which converts the model into a series of thin layers and produces a G-code file containing instructions tailored to a specific type of 3D printer.

Slicers dissect your digital model into layers, determine the best path to lay down each layer of filament, and program the rate at which material is ejected from the nozzle. This process is critical for the successful creation of your 3D print.

Popular slicing software options are Cura and Slic3r – both are compatible with most 3D printers and offer free editions.

11.5. Your First Print

It's time to send your first project to the 3D printer. Make sure the printer is appropriately calibrated. Calibration can vary with 3D printer brands, so be sure to follow the instructions that come with your specific machine. Remember to check the selected material, printing speed, and layer resolution in the slicer settings. These parameters can dramatically affect your print quality.

With everything ready, you can start your print! Your 3D printer should slowly build your object layer by layer until it's complete. This process could take a couple of hours to more than a day, depending on the complexity and size of the design.

If you find that your finished product isn't quite what you expected, don't be disheartened. Many details could cause a print to fail or not turn out as anticipated, ranging from settings errors to a misplaced spool of filament. With each misstep, you'll have learned another detail to watch for, and your subsequent attempts should run smoother.

11.6. Next Steps

Once you have created your first 3D printed object, the sky's the limit! You could take on more complex projects, experiment with different materials, or even refine your 3D modeling skills. Embrace the failures and the successes, don't cease to learn, innovate, and test the limits of your creativity.

Starting with a general understanding of the process, choosing the right printer, learning to use modeling software, to creating your first 3D print - each facet represents a step in your journey through 3D printing. By understanding each stage and mastering them one by one, you'll be converting your ideas into actual tangible objects in no time.